MATH
Multiplication

THIS BOOK BELONG TO

....

$$6 \times 7$$

$$7 \times 2$$

$$6 \times 4$$

$$8 \times 6$$

$$8 \times 4$$

$$6 \times 9$$

$$2 \times 5$$

$$8 \times 8$$

$$3 \times 1$$

$$1 \times 9$$

$$5 \times 2$$

$$3 \times 7$$

$$7 \times 5$$

$$4 \times 7$$

$$2 \times 4$$

$$5 \times 1$$

$$5 \times 3$$

$$9 \times 9$$

$$7 \times 7$$

$$3 \times 8$$

$$6 \times 8$$

$$4 \times 10$$

$$9 \times 7$$

$$4 \times 6$$

$$7 \times 6$$

$$1 \times 2$$

$$9 \times 6$$

$$2 \times 2$$

$$1 \times 7$$

$$9 \times 5$$

4 × 3	5 × 8	8 × 9
4 × 8	3 × 3	7 × 4
10 × 1	9 × 8	5 × 6
9 × 2	2 × 1	6 × 6
2 × 3	4 × 1	3 × 4

$$
\begin{array}{r} 2 \\ \times\ 9 \\ \hline \end{array}
\qquad
\begin{array}{r} 10 \\ \times\ 2 \\ \hline \end{array}
\qquad
\begin{array}{r} 8 \\ \times\ 1 \\ \hline \end{array}
$$

$$
\begin{array}{r} 10 \\ \times\ 5 \\ \hline \end{array}
\qquad
\begin{array}{r} 3 \\ \times\ 10 \\ \hline \end{array}
\qquad
\begin{array}{r} 3 \\ \times\ 5 \\ \hline \end{array}
$$

$$
\begin{array}{r} 9 \\ \times\ 3 \\ \hline \end{array}
\qquad
\begin{array}{r} 5 \\ \times\ 7 \\ \hline \end{array}
\qquad
\begin{array}{r} 10 \\ \times\ 9 \\ \hline \end{array}
$$

$$
\begin{array}{r} 7 \\ \times\ 8 \\ \hline \end{array}
\qquad
\begin{array}{r} 10 \\ \times\ 6 \\ \hline \end{array}
\qquad
\begin{array}{r} 1 \\ \times\ 4 \\ \hline \end{array}
$$

$$
\begin{array}{r} 6 \\ \times\ 3 \\ \hline \end{array}
\qquad
\begin{array}{r} 7 \\ \times\ 9 \\ \hline \end{array}
\qquad
\begin{array}{r} 9 \\ \times\ 1 \\ \hline \end{array}
$$

9	6	10
× 4	× 5	× 7

1	2	8
× 3	× 10	× 7

5	4	1
× 5	× 2	× 6

7	3	5
× 1	× 6	× 9

5	8	6
× 10	× 5	× 2

2 × 7	8 × 10	6 × 10
2 × 8	4 × 4	9 × 10
7 × 3	10 × 4	4 × 9
8 × 3	5 × 4	7 × 10
4 × 5	6 × 1	3 × 9

$$6 \times 4$$

$$6 \times 5$$

$$8 \times 6$$

$$4 \times 9$$

$$6 \times 6$$

$$10 \times 3$$

$$2 \times 2$$

$$8 \times 8$$

$$5 \times 8$$

$$5 \times 9$$

$$7 \times 6$$

$$1 \times 7$$

$$10 \times 5$$

$$4 \times 2$$

$$4 \times 8$$

multiplication

$$\begin{array}{r} 10 \\ \times\ 4 \\ \hline \end{array}$$

....

$$\begin{array}{r} 9 \\ \times\ 7 \\ \hline \end{array}$$

....

$$\begin{array}{r} 4 \\ \times\ 3 \\ \hline \end{array}$$

....

$$\begin{array}{r} 1 \\ \times\ 1 \\ \hline \end{array}$$

....

$$\begin{array}{r} 9 \\ \times\ 2 \\ \hline \end{array}$$

....

$$\begin{array}{r} 7 \\ \times\ 1 \\ \hline \end{array}$$

....

$$\begin{array}{r} 10 \\ \times\ 7 \\ \hline \end{array}$$

....

$$\begin{array}{r} 2 \\ \times\ 1 \\ \hline \end{array}$$

....

$$\begin{array}{r} 10 \\ \times\ 1 \\ \hline \end{array}$$

....

$$\begin{array}{r} 4 \\ \times\ 7 \\ \hline \end{array}$$

....

$$\begin{array}{r} 7 \\ \times\ 2 \\ \hline \end{array}$$

....

$$\begin{array}{r} 5 \\ \times\ 2 \\ \hline \end{array}$$

....

$$\begin{array}{r} 6 \\ \times\ 9 \\ \hline \end{array}$$

....

$$\begin{array}{r} 5 \\ \times\ 5 \\ \hline \end{array}$$

....

$$\begin{array}{r} 3 \\ \times\ 3 \\ \hline \end{array}$$

....

multiplication

3 × 8	3 × 1	4 × 6
8 × 5	9 × 4	3 × 6
7 × 10	3 × 5	6 × 1
8 × 9	1 × 3	7 × 3
5 × 3	8 × 4	2 × 10

5 $\times$ 7	9 $\times$ 3	5 $\times$ 4
8 $\times$ 2	10 $\times$ 2	10 $\times$ 8
3 $\times$ 9	8 $\times$ 1	3 $\times$ 2
4 $\times$ 5	2 $\times$ 7	4 $\times$ 1
5 $\times$ 10	6 $\times$ 2	6 $\times$ 8

9 × 6	7 × 5	7 × 4
7 × 7	4 × 4	2 × 4
2 × 9	7 × 9	6 × 3
2 × 6	2 × 3	1 × 4
8 × 10	3 × 4	6 × 7

8 × 3	3 × 7	10 × 9
9 × 5	2 × 5	8 × 7
1 × 5	10 × 6	1 × 8
9 × 9	1 × 10	9 × 10
9 × 8	3 × 10	6 × 10

1 × 5	3 × 7	6 × 9
2 × 9	1 × 7	6 × 10
2 × 2	9 × 5	8 × 2
7 × 10	4 × 10	4 × 6
1 × 9	6 × 1	7 × 3

$$9 \times 8 = $$

$$8 \times 8 = $$

$$2 \times 8 = $$

$$8 \times 1 = $$

$$9 \times 4 = $$

$$3 \times 3 = $$

$$7 \times 4 = $$

$$4 \times 3 = $$

$$7 \times 7 = $$

$$5 \times 6 = $$

$$9 \times 1 = $$

$$2 \times 4 = $$

$$4 \times 1 = $$

$$8 \times 3 = $$

$$5 \times 2 = $$

multiplication

$$
\begin{array}{r} 3 \\ \times\ 10 \\ \hline \end{array}
\qquad
\begin{array}{r} 3 \\ \times\ 4 \\ \hline \end{array}
\qquad
\begin{array}{r} 10 \\ \times\ 7 \\ \hline \end{array}
$$

$$
\begin{array}{r} 9 \\ \times\ 7 \\ \hline \end{array}
\qquad
\begin{array}{r} 6 \\ \times\ 4 \\ \hline \end{array}
\qquad
\begin{array}{r} 7 \\ \times\ 9 \\ \hline \end{array}
$$

$$
\begin{array}{r} 5 \\ \times\ 8 \\ \hline \end{array}
\qquad
\begin{array}{r} 10 \\ \times\ 6 \\ \hline \end{array}
\qquad
\begin{array}{r} 5 \\ \times\ 9 \\ \hline \end{array}
$$

$$
\begin{array}{r} 9 \\ \times\ 3 \\ \hline \end{array}
\qquad
\begin{array}{r} 10 \\ \times\ 8 \\ \hline \end{array}
\qquad
\begin{array}{r} 1 \\ \times\ 3 \\ \hline \end{array}
$$

$$
\begin{array}{r} 6 \\ \times\ 3 \\ \hline \end{array}
\qquad
\begin{array}{r} 3 \\ \times\ 2 \\ \hline \end{array}
\qquad
\begin{array}{r} 7 \\ \times\ 8 \\ \hline \end{array}
$$

multiplication

$$4 \times 7$$

$$5 \times 4$$

$$5 \times 5$$

$$10 \times 2$$

$$8 \times 4$$

$$7 \times 6$$

$$7 \times 5$$

$$3 \times 6$$

$$8 \times 6$$

$$3 \times 9$$

$$4 \times 4$$

$$6 \times 2$$

$$3 \times 1$$

$$2 \times 6$$

$$1 \times 1$$

$$\begin{array}{r} 1 \\ \times\ 4 \\ \hline \end{array}$$
$$\begin{array}{r} 8 \\ \times\ 5 \\ \hline \end{array}$$
$$\begin{array}{r} 9 \\ \times\ 9 \\ \hline \end{array}$$

$$\begin{array}{r} 9 \\ \times\ 6 \\ \hline \end{array}$$
$$\begin{array}{r} 6 \\ \times\ 8 \\ \hline \end{array}$$
$$\begin{array}{r} 8 \\ \times\ 9 \\ \hline \end{array}$$

$$\begin{array}{r} 3 \\ \times\ 5 \\ \hline \end{array}$$
$$\begin{array}{r} 3 \\ \times\ 8 \\ \hline \end{array}$$
$$\begin{array}{r} 2 \\ \times\ 3 \\ \hline \end{array}$$

$$\begin{array}{r} 8 \\ \times\ 7 \\ \hline \end{array}$$
$$\begin{array}{r} 2 \\ \times\ 7 \\ \hline \end{array}$$
$$\begin{array}{r} 5 \\ \times\ 10 \\ \hline \end{array}$$

$$\begin{array}{r} 5 \\ \times\ 7 \\ \hline \end{array}$$
$$\begin{array}{r} 5 \\ \times\ 3 \\ \hline \end{array}$$
$$\begin{array}{r} 2 \\ \times\ 5 \\ \hline \end{array}$$

6	6	6
× 6	× 7	× 5

10	4	2
× 9	× 9	× 1

2	9	4
× 10	× 2	× 2

5	10	7
× 1	× 4	× 2

7	1	8
× 1	× 2	× 10

$$8 \times 4$$

$$1 \times 5$$

$$4 \times 7$$

....

$$9 \times 8$$

$$2 \times 6$$

$$5 \times 2$$

....

$$3 \times 4$$

$$4 \times 2$$

$$5 \times 3$$

....

$$3 \times 6$$

$$7 \times 6$$

$$9 \times 5$$

....

$$4 \times 9$$

$$10 \times 7$$

$$9 \times 6$$

....

6 × 6	4 × 6	9 × 3
8 × 2	5 × 9	3 × 5
2 × 3	2 × 7	8 × 8
8 × 5	5 × 4	6 × 8
5 × 8	6 × 9	10 × 8

5 × 5	7 × 8	10 × 2
5 × 7	8 × 6	7 × 7
2 × 9	7 × 2	6 × 7
10 × 5	8 × 3	2 × 10
4 × 8	1 × 4	2 × 4

3	2	10
× 3	× 8	× 3

4	7	1
× 1	× 3	× 10

2	4	9
× 5	× 5	× 2

1	9	3
× 6	× 4	× 9

6	6	3
× 10	× 4	× 7

5 × 1	10 × 9	1 × 2
8 × 9	7 × 4	1 × 9
4 × 4	6 × 5	3 × 1
2 × 2	7 × 9	8 × 7
7 × 5	8 × 1	10 × 1

$$\begin{array}{r} 10 \\ \times\ 6 \\ \hline \end{array}$$

$$\begin{array}{r} 10 \\ \times\ 4 \\ \hline \end{array}$$

$$\begin{array}{r} 6 \\ \times\ 3 \\ \hline \end{array}$$

$$\begin{array}{r} 5 \\ \times\ 6 \\ \hline \end{array}$$

$$\begin{array}{r} 1 \\ \times\ 8 \\ \hline \end{array}$$

$$\begin{array}{r} 6 \\ \times\ 1 \\ \hline \end{array}$$

$$\begin{array}{r} 6 \\ \times\ 2 \\ \hline \end{array}$$

$$\begin{array}{r} 3 \\ \times\ 10 \\ \hline \end{array}$$

$$\begin{array}{r} 1 \\ \times\ 3 \\ \hline \end{array}$$

$$\begin{array}{r} 3 \\ \times\ 8 \\ \hline \end{array}$$

$$\begin{array}{r} 9 \\ \times\ 7 \\ \hline \end{array}$$

$$\begin{array}{r} 9 \\ \times\ 9 \\ \hline \end{array}$$

$$\begin{array}{r} 1 \\ \times\ 7 \\ \hline \end{array}$$

$$\begin{array}{r} 2 \\ \times\ 1 \\ \hline \end{array}$$

$$\begin{array}{r} 10 \\ \times\ 10 \\ \hline \end{array}$$

3 × 3	3 × 4	5 × 8
4 × 10	9 × 8	6 × 7
9 × 7	1 × 5	8 × 2
6 × 10	9 × 2	2 × 8
6 × 3	8 × 5	8 × 6

4 × 3	7 × 9	7 × 4
8 × 8	8 × 1	6 × 6
9 × 9	4 × 2	3 × 7
2 × 5	2 × 10	9 × 4
2 × 6	5 × 4	2 × 2

multiplication

$$\begin{array}{r} 4 \\ \times\ 5 \\ \hline \end{array}$$

$$\begin{array}{r} 9 \\ \times\ 3 \\ \hline \end{array}$$

$$\begin{array}{r} 10 \\ \times\ 3 \\ \hline \end{array}$$

$$\begin{array}{r} 5 \\ \times\ 3 \\ \hline \end{array}$$

$$\begin{array}{r} 10 \\ \times\ 10 \\ \hline \end{array}$$

$$\begin{array}{r} 3 \\ \times\ 5 \\ \hline \end{array}$$

$$\begin{array}{r} 10 \\ \times\ 6 \\ \hline \end{array}$$

$$\begin{array}{r} 9 \\ \times\ 1 \\ \hline \end{array}$$

$$\begin{array}{r} 4 \\ \times\ 4 \\ \hline \end{array}$$

$$\begin{array}{r} 2 \\ \times\ 3 \\ \hline \end{array}$$

$$\begin{array}{r} 3 \\ \times\ 9 \\ \hline \end{array}$$

$$\begin{array}{r} 5 \\ \times\ 9 \\ \hline \end{array}$$

$$\begin{array}{r} 4 \\ \times\ 1 \\ \hline \end{array}$$

$$\begin{array}{r} 8 \\ \times\ 9 \\ \hline \end{array}$$

$$\begin{array}{r} 6 \\ \times\ 5 \\ \hline \end{array}$$

multiplication

$$\begin{array}{r} 5 \\ \times\ 10 \\ \hline \end{array}$$

$$\begin{array}{r} 2 \\ \times\ 4 \\ \hline \end{array}$$

$$\begin{array}{r} 8 \\ \times\ 7 \\ \hline \end{array}$$

$$\begin{array}{r} 6 \\ \times\ 8 \\ \hline \end{array}$$

$$\begin{array}{r} 8 \\ \times\ 10 \\ \hline \end{array}$$

$$\begin{array}{r} 5 \\ \times\ 7 \\ \hline \end{array}$$

$$\begin{array}{r} 8 \\ \times\ 4 \\ \hline \end{array}$$

$$\begin{array}{r} 7 \\ \times\ 5 \\ \hline \end{array}$$

$$\begin{array}{r} 9 \\ \times\ 5 \\ \hline \end{array}$$

$$\begin{array}{r} 7 \\ \times\ 2 \\ \hline \end{array}$$

$$\begin{array}{r} 9 \\ \times\ 10 \\ \hline \end{array}$$

$$\begin{array}{r} 5 \\ \times\ 5 \\ \hline \end{array}$$

$$\begin{array}{r} 6 \\ \times\ 9 \\ \hline \end{array}$$

$$\begin{array}{r} 1 \\ \times\ 9 \\ \hline \end{array}$$

$$\begin{array}{r} 5 \\ \times\ 2 \\ \hline \end{array}$$

multiplication

6 × 2	3 × 2	5 × 6
4 × 8	9 × 6	7 × 6
6 × 4	4 × 7	4 × 9
7 × 3	2 × 1	10 × 7
8 × 3	1 × 1	5 × 1

$$3 \times 6$$

$$2 \times 9$$

$$4 \times 6$$

$$1 \times 2$$

$$1 \times 7$$

$$2 \times 7$$

$$1 \times 8$$

$$10 \times 2$$

$$3 \times 1$$

$$6 \times 1$$

$$7 \times 8$$

$$10 \times 8$$

$$3 \times 8$$

$$1 \times 3$$

$$10 \times 1$$

8	4	1
× 10	× 1	× 1
5	4	2
× 10	× 6	× 1
9	4	8
× 6	× 10	× 3
9	8	6
× 9	× 1	× 2
8	1	2
× 4	× 9	× 8

$$\begin{array}{r} 8 \\ \times\ 6 \\ \hline \end{array}$$

$$\begin{array}{r} 2 \\ \times\ 4 \\ \hline \end{array}$$

$$\begin{array}{r} 9 \\ \times\ 7 \\ \hline \end{array}$$

$$\begin{array}{r} 2 \\ \times\ 7 \\ \hline \end{array}$$

$$\begin{array}{r} 6 \\ \times\ 6 \\ \hline \end{array}$$

$$\begin{array}{r} 7 \\ \times\ 8 \\ \hline \end{array}$$

$$\begin{array}{r} 9 \\ \times\ 2 \\ \hline \end{array}$$

$$\begin{array}{r} 1 \\ \times\ 6 \\ \hline \end{array}$$

$$\begin{array}{r} 4 \\ \times\ 5 \\ \hline \end{array}$$

$$\begin{array}{r} 6 \\ \times\ 8 \\ \hline \end{array}$$

$$\begin{array}{r} 4 \\ \times\ 8 \\ \hline \end{array}$$

$$\begin{array}{r} 5 \\ \times\ 3 \\ \hline \end{array}$$

$$\begin{array}{r} 3 \\ \times\ 7 \\ \hline \end{array}$$

$$\begin{array}{r} 10 \\ \times\ 6 \\ \hline \end{array}$$

$$\begin{array}{r} 3 \\ \times\ 1 \\ \hline \end{array}$$

$$7 \times 10$$

$$8 \times 5$$

$$5 \times 4$$

....

....

....

$$1 \times 2$$

$$3 \times 9$$

$$5 \times 8$$

....

....

....

$$9 \times 3$$

$$7 \times 4$$

$$4 \times 4$$

....

....

....

$$8 \times 2$$

$$7 \times 3$$

$$6 \times 5$$

....

....

....

$$9 \times 5$$

$$10 \times 1$$

$$2 \times 6$$

....

....

....

5	6	6
× 7	× 7	× 10

7	7	7
× 9	× 1	× 7

6	10	5
× 9	× 10	× 5

2	8	5
× 3	× 8	× 6

3	2	10
× 5	× 9	× 7

10	5	7
× 2	× 2	× 2
....	....	....
4	1	6
× 3	× 5	× 1
....	....	....
2	2	9
× 2	× 5	× 8
....	....	....
8	3	4
× 9	× 8	× 7
....	....	....
3	3	8
× 2	× 6	× 7
....	....	....

$$7 \times 6$$

$$9 \times 4$$

$$3 \times 10$$

$$3 \times 4$$

$$1 \times 7$$

$$4 \times 9$$

$$4 \times 2$$

$$6 \times 4$$

$$9 \times 1$$

$$1 \times 10$$

$$7 \times 5$$

$$6 \times 3$$

$$3 \times 3$$

$$10 \times 3$$

$$2 \times 10$$

7 × 10	6 × 2	2 × 4
4 × 6	2 × 6	6 × 4
3 × 2	4 × 3	5 × 5
4 × 10	9 × 7	3 × 6
6 × 8	10 × 1	2 × 3

3 × 4	6 × 9	9 × 4
9 × 9	9 × 8	9 × 6
5 × 9	2 × 10	5 × 3
9 × 1	6 × 7	5 × 10
8 × 1	4 × 4	8 × 4

$$3 \times 1$$

$$7 \times 6$$

$$6 \times 5$$

$$8 \times 7$$

$$2 \times 7$$

$$7 \times 1$$

$$7 \times 9$$

$$8 \times 9$$

$$4 \times 1$$

$$3 \times 3$$

$$3 \times 8$$

$$7 \times 5$$

$$6 \times 3$$

$$5 \times 2$$

$$2 \times 2$$

7	8	5
× 4	× 6	× 8

4	8	1
× 2	× 8	× 5

5	10	4
× 4	× 6	× 5

5	6	8
× 7	× 1	× 2

10	5	9
× 9	× 1	× 3

8 × 3	1 × 9	2 × 5
9 × 2	4 × 7	1 × 2
7 × 3	1 × 4	3 × 7
3 × 9	7 × 8	10 × 4
2 × 8	7 × 7	5 × 6

8 × 5	6 × 6	10 × 3
2 × 1	10 × 2	2 × 9
9 × 10	4 × 8	7 × 2
3 × 5	1 × 8	4 × 9
6 × 10	10 × 10	9 × 5

2 × 4	4 × 9	5 × 4
2 × 3	5 × 6	9 × 3
1 × 7	2 × 7	6 × 7
5 × 3	2 × 10	2 × 8
4 × 10	1 × 3	3 × 9

$$8 \times 9$$

$$3 \times 6$$

$$8 \times 3$$

$$4 \times 5$$

$$3 \times 8$$

$$9 \times 8$$

$$7 \times 4$$

$$10 \times 7$$

$$6 \times 2$$

$$3 \times 5$$

$$6 \times 5$$

$$3 \times 3$$

$$10 \times 2$$

$$2 \times 9$$

$$6 \times 10$$

5 × 9	7 × 7	1 × 2
8 × 4	5 × 8	5 × 5
4 × 7	7 × 1	10 × 3
3 × 2	9 × 7	6 × 9
9 × 1	3 × 7	7 × 9

$$\begin{array}{r} 8 \\ \times\ 6 \\ \hline \end{array}\qquad \begin{array}{r} 2 \\ \times\ 5 \\ \hline \end{array}\qquad \begin{array}{r} 8 \\ \times\ 8 \\ \hline \end{array}$$

$$\begin{array}{r} 7 \\ \times\ 5 \\ \hline \end{array}\qquad \begin{array}{r} 9 \\ \times\ 6 \\ \hline \end{array}\qquad \begin{array}{r} 8 \\ \times\ 7 \\ \hline \end{array}$$

$$\begin{array}{r} 10 \\ \times\ 5 \\ \hline \end{array}\qquad \begin{array}{r} 6 \\ \times\ 6 \\ \hline \end{array}\qquad \begin{array}{r} 7 \\ \times\ 10 \\ \hline \end{array}$$

$$\begin{array}{r} 4 \\ \times\ 4 \\ \hline \end{array}\qquad \begin{array}{r} 2 \\ \times\ 1 \\ \hline \end{array}\qquad \begin{array}{r} 4 \\ \times\ 8 \\ \hline \end{array}$$

$$\begin{array}{r} 7 \\ \times\ 2 \\ \hline \end{array}\qquad \begin{array}{r} 4 \\ \times\ 2 \\ \hline \end{array}\qquad \begin{array}{r} 1 \\ \times\ 5 \\ \hline \end{array}$$

8 × 10	6 × 1	9 × 9
7 × 3	8 × 5	7 × 6
10 × 10	10 × 8	9 × 5
8 × 2	7 × 8	6 × 3
2 × 2	3 × 4	4 × 6

5 × 1	9 × 10	1 × 9
4 × 1	9 × 4	10 × 9
6 × 8	8 × 1	6 × 4
3 × 10	5 × 2	5 × 7
10 × 4	10 × 1	9 × 2

8 × 6	7 × 4	1 × 10
2 × 5	4 × 10	5 × 5
9 × 5	6 × 7	6 × 8
5 × 8	4 × 6	10 × 4
6 × 10	3 × 3	3 × 6

$$6 \times 4$$

$$2 \times 3$$

$$7 \times 5$$

$$6 \times 5$$

$$7 \times 3$$

$$2 \times 1$$

$$6 \times 3$$

$$7 \times 2$$

$$3 \times 2$$

$$3 \times 4$$

$$4 \times 7$$

$$5 \times 2$$

$$5 \times 7$$

$$2 \times 9$$

$$1 \times 3$$

multiplication

$\begin{array}{r} 10 \\ \times\ 6 \\ \hline \end{array}$	$\begin{array}{r} 7 \\ \times\ 8 \\ \hline \end{array}$	$\begin{array}{r} 5 \\ \times\ 10 \\ \hline \end{array}$
$\begin{array}{r} 6 \\ \times\ 2 \\ \hline \end{array}$	$\begin{array}{r} 4 \\ \times\ 1 \\ \hline \end{array}$	$\begin{array}{r} 1 \\ \times\ 5 \\ \hline \end{array}$
$\begin{array}{r} 9 \\ \times\ 6 \\ \hline \end{array}$	$\begin{array}{r} 3 \\ \times\ 5 \\ \hline \end{array}$	$\begin{array}{r} 10 \\ \times\ 1 \\ \hline \end{array}$
$\begin{array}{r} 10 \\ \times\ 8 \\ \hline \end{array}$	$\begin{array}{r} 9 \\ \times\ 7 \\ \hline \end{array}$	$\begin{array}{r} 5 \\ \times\ 1 \\ \hline \end{array}$
$\begin{array}{r} 7 \\ \times\ 7 \\ \hline \end{array}$	$\begin{array}{r} 5 \\ \times\ 9 \\ \hline \end{array}$	$\begin{array}{r} 2 \\ \times\ 2 \\ \hline \end{array}$

1 × 2	9 × 9	7 × 6
8 × 5	8 × 3	1 × 6
2 × 10	7 × 1	9 × 1
10 × 9	8 × 8	8 × 4
4 × 4	8 × 9	3 × 9

4 $\times$ 3	1 $\times$ 1	4 $\times$ 8
2 $\times$ 4	6 $\times$ 6	9 $\times$ 8
8 $\times$ 1	9 $\times$ 2	3 $\times$ 10
3 $\times$ 7	5 $\times$ 4	8 $\times$ 2
5 $\times$ 6	10 $\times$ 5	2 $\times$ 8

4	9	7
× 9	× 3	× 10

10	5	2
× 2	× 3	× 7

10	4	9
× 3	× 5	× 4

8	3	7
× 10	× 8	× 9

10	2	4
× 7	× 6	× 2

3	9	1
× 2	× 7	× 9

1	4	8
× 3	× 7	× 2

6	8	2
× 5	× 5	× 1

10	7	5
× 3	× 7	× 6

1	2	8
× 5	× 4	× 10

$$\begin{array}{r} 8 \\ \times\ 6 \\ \hline \end{array}$$
....

$$\begin{array}{r} 7 \\ \times\ 9 \\ \hline \end{array}$$
....

$$\begin{array}{r} 2 \\ \times\ 7 \\ \hline \end{array}$$
....

$$\begin{array}{r} 3 \\ \times\ 4 \\ \hline \end{array}$$
....

$$\begin{array}{r} 4 \\ \times\ 9 \\ \hline \end{array}$$
....

$$\begin{array}{r} 4 \\ \times\ 2 \\ \hline \end{array}$$
....

$$\begin{array}{r} 10 \\ \times\ 6 \\ \hline \end{array}$$
....

$$\begin{array}{r} 3 \\ \times\ 5 \\ \hline \end{array}$$
....

$$\begin{array}{r} 1 \\ \times\ 6 \\ \hline \end{array}$$
....

$$\begin{array}{r} 5 \\ \times\ 7 \\ \hline \end{array}$$
....

$$\begin{array}{r} 3 \\ \times\ 6 \\ \hline \end{array}$$
....

$$\begin{array}{r} 7 \\ \times\ 4 \\ \hline \end{array}$$
....

$$\begin{array}{r} 7 \\ \times\ 6 \\ \hline \end{array}$$
....

$$\begin{array}{r} 3 \\ \times\ 10 \\ \hline \end{array}$$
....

$$\begin{array}{r} 9 \\ \times\ 8 \\ \hline \end{array}$$
....

multiplication

$$4 \times 10$$

$$8 \times 9$$

$$7 \times 2$$

$$9 \times 4$$

$$1 \times 4$$

$$4 \times 5$$

$$3 \times 9$$

$$5 \times 2$$

$$3 \times 7$$

$$4 \times 3$$

$$6 \times 2$$

$$4 \times 6$$

$$8 \times 1$$

$$8 \times 4$$

$$3 \times 3$$

$$\begin{array}{r} 2 \\ \times\ 3 \\ \hline \end{array}$$

$$\begin{array}{r} 10 \\ \times\ 4 \\ \hline \end{array}$$

$$\begin{array}{r} 2 \\ \times\ 10 \\ \hline \end{array}$$

$$\begin{array}{r} 5 \\ \times\ 5 \\ \hline \end{array}$$

$$\begin{array}{r} 10 \\ \times\ 1 \\ \hline \end{array}$$

$$\begin{array}{r} 7 \\ \times\ 5 \\ \hline \end{array}$$

$$\begin{array}{r} 8 \\ \times\ 3 \\ \hline \end{array}$$

$$\begin{array}{r} 6 \\ \times\ 7 \\ \hline \end{array}$$

$$\begin{array}{r} 9 \\ \times\ 5 \\ \hline \end{array}$$

$$\begin{array}{r} 7 \\ \times\ 3 \\ \hline \end{array}$$

$$\begin{array}{r} 2 \\ \times\ 8 \\ \hline \end{array}$$

$$\begin{array}{r} 1 \\ \times\ 7 \\ \hline \end{array}$$

$$\begin{array}{r} 4 \\ \times\ 4 \\ \hline \end{array}$$

$$\begin{array}{r} 2 \\ \times\ 2 \\ \hline \end{array}$$

$$\begin{array}{r} 7 \\ \times\ 8 \\ \hline \end{array}$$

3	6	6
× 8	× 10	× 4
....	....	....

2	6	6
× 5	× 6	× 9
....	....	....

5	9	5
× 9	× 10	× 8
....	....	....

5	9	1
× 4	× 2	× 1
....	....	....

10	2	4
× 8	× 9	× 8
....	....	....

$$5 \times 10$$

$$7 \times 10$$

$$4 \times 1$$

$$6 \times 8$$

$$9 \times 6$$

$$5 \times 3$$

$$9 \times 9$$

$$10 \times 9$$

$$9 \times 3$$

$$10 \times 10$$

$$2 \times 6$$

$$8 \times 7$$

$$5 \times 1$$

$$1 \times 10$$

$$10 \times 5$$

$$
\begin{array}{r}
5 \\
\times\ 7 \\
\hline
\end{array}
\qquad
\begin{array}{r}
4 \\
\times\ 1 \\
\hline
\end{array}
\qquad
\begin{array}{r}
7 \\
\times\ 7 \\
\hline
\end{array}
$$

$$
\begin{array}{r}
3 \\
\times\ 2 \\
\hline
\end{array}
\qquad
\begin{array}{r}
3 \\
\times\ 5 \\
\hline
\end{array}
\qquad
\begin{array}{r}
4 \\
\times\ 6 \\
\hline
\end{array}
$$

$$
\begin{array}{r}
6 \\
\times\ 7 \\
\hline
\end{array}
\qquad
\begin{array}{r}
2 \\
\times\ 10 \\
\hline
\end{array}
\qquad
\begin{array}{r}
3 \\
\times\ 8 \\
\hline
\end{array}
$$

$$
\begin{array}{r}
6 \\
\times\ 8 \\
\hline
\end{array}
\qquad
\begin{array}{r}
8 \\
\times\ 7 \\
\hline
\end{array}
\qquad
\begin{array}{r}
2 \\
\times\ 4 \\
\hline
\end{array}
$$

$$
\begin{array}{r}
6 \\
\times\ 9 \\
\hline
\end{array}
\qquad
\begin{array}{r}
2 \\
\times\ 7 \\
\hline
\end{array}
\qquad
\begin{array}{r}
1 \\
\times\ 4 \\
\hline
\end{array}
$$

2	7	8
× 3	× 3	× 6

6	5	10
× 1	× 8	× 10

8	6	7
× 4	× 6	× 6

4	8	1
× 10	× 8	× 1

4	8	9
× 7	× 9	× 8

$$3 \times 4$$

$$5 \times 10$$

$$4 \times 2$$

$$7 \times 8$$

$$1 \times 7$$

$$10 \times 3$$

$$6 \times 3$$

$$9 \times 5$$

$$5 \times 6$$

$$2 \times 8$$

$$2 \times 2$$

$$5 \times 1$$

$$2 \times 5$$

$$6 \times 5$$

$$9 \times 4$$

8 × 3	4 × 8	3 × 9
3 × 6	4 × 3	1 × 6
5 × 3	7 × 2	3 × 7
5 × 4	9 × 2	4 × 9
9 × 6	7 × 4	9 × 7

5 × 9	7 × 9	8 × 2
4 × 5	3 × 3	8 × 5
7 × 10	8 × 1	10 × 1
1 × 10	8 × 10	6 × 4
5 × 5	5 × 2	9 × 9

1 × 9	2 × 9	9 × 3
6 × 2	10 × 6	9 × 1
6 × 10	4 × 4	9 × 10
7 × 5	1 × 2	10 × 5
2 × 6	3 × 1	1 × 8

3 × 9	9 × 8	4 × 7
9 × 5	9 × 2	1 × 5
8 × 5	3 × 6	9 × 1
10 × 7	3 × 1	2 × 4
3 × 5	5 × 7	5 × 10

3	4	9
× 4	× 4	× 3

8	2	7
× 2	× 5	× 9

5	8	7
× 2	× 4	× 6

10	7	6
× 1	× 4	× 2

8	6	3
× 7	× 9	× 3

multiplication

1 × 9	9 × 7	2 × 6
7 × 7	4 × 8	5 × 8
4 × 2	7 × 2	10 × 9
10 × 3	8 × 1	8 × 3
10 × 10	1 × 7	10 × 4

$$3 \times 2$$

$$9 \times 4$$

$$8 \times 8$$

$$2 \times 2$$

$$10 \times 2$$

$$9 \times 6$$

$$5 \times 6$$

$$2 \times 3$$

$$9 \times 9$$

$$5 \times 9$$

$$6 \times 4$$

$$8 \times 9$$

$$4 \times 6$$

$$6 \times 5$$

$$3 \times 8$$

$$6 \times 6 = $$

$$6 \times 1 = $$

$$7 \times 8 = $$

$$8 \times 6 = $$

$$2 \times 1 = $$

$$7 \times 5 = $$

$$2 \times 7 = $$

$$6 \times 7 = $$

$$10 \times 8 = $$

$$1 \times 1 = $$

$$4 \times 3 = $$

$$2 \times 9 = $$

$$7 \times 10 = $$

$$5 \times 5 = $$

$$1 \times 3 = $$

1	6	7
× 2	× 10	× 1

2	4	3
× 8	× 5	× 7

5	5	10
× 3	× 4	× 5

6	7	1
× 3	× 3	× 4

6	1	2
× 8	× 8	× 10

multiplication

$$\begin{array}{r} 1 \\ \times\ 8 \\ \hline \end{array}$$

$$\begin{array}{r} 3 \\ \times\ 3 \\ \hline \end{array}$$

$$\begin{array}{r} 7 \\ \times\ 7 \\ \hline \end{array}$$

$$\begin{array}{r} 6 \\ \times\ 9 \\ \hline \end{array}$$

$$\begin{array}{r} 9 \\ \times\ 1 \\ \hline \end{array}$$

$$\begin{array}{r} 4 \\ \times\ 5 \\ \hline \end{array}$$

$$\begin{array}{r} 3 \\ \times\ 5 \\ \hline \end{array}$$

$$\begin{array}{r} 7 \\ \times\ 9 \\ \hline \end{array}$$

$$\begin{array}{r} 6 \\ \times\ 8 \\ \hline \end{array}$$

$$\begin{array}{r} 4 \\ \times\ 10 \\ \hline \end{array}$$

$$\begin{array}{r} 4 \\ \times\ 4 \\ \hline \end{array}$$

$$\begin{array}{r} 4 \\ \times\ 2 \\ \hline \end{array}$$

$$\begin{array}{r} 7 \\ \times\ 2 \\ \hline \end{array}$$

$$\begin{array}{r} 9 \\ \times\ 4 \\ \hline \end{array}$$

$$\begin{array}{r} 3 \\ \times\ 4 \\ \hline \end{array}$$

5 × 8	3 × 6	2 × 4
6 × 10	10 × 6	4 × 8
3 × 7	5 × 3	9 × 6
1 × 5	2 × 1	5 × 5
6 × 7	7 × 1	5 × 4

$$8 \times 9$$

$$9 \times 10$$

$$9 \times 2$$

$$7 \times 4$$

$$3 \times 2$$

$$2 \times 7$$

$$4 \times 1$$

$$6 \times 5$$

$$7 \times 3$$

$$6 \times 2$$

$$8 \times 3$$

$$6 \times 4$$

$$2 \times 3$$

$$5 \times 10$$

$$6 \times 1$$

$$\begin{array}{r} 8 \\ \times\ 7 \\ \hline \end{array}$$ ....

$$\begin{array}{r} 2 \\ \times\ 8 \\ \hline \end{array}$$ ....

$$\begin{array}{r} 5 \\ \times\ 6 \\ \hline \end{array}$$ ....

$$\begin{array}{r} 8 \\ \times\ 2 \\ \hline \end{array}$$ ....

$$\begin{array}{r} 4 \\ \times\ 3 \\ \hline \end{array}$$ ....

$$\begin{array}{r} 5 \\ \times\ 7 \\ \hline \end{array}$$ ....

$$\begin{array}{r} 8 \\ \times\ 6 \\ \hline \end{array}$$ ....

$$\begin{array}{r} 6 \\ \times\ 3 \\ \hline \end{array}$$ ....

$$\begin{array}{r} 1 \\ \times\ 7 \\ \hline \end{array}$$ ....

$$\begin{array}{r} 7 \\ \times\ 8 \\ \hline \end{array}$$ ....

$$\begin{array}{r} 9 \\ \times\ 5 \\ \hline \end{array}$$ ....

$$\begin{array}{r} 10 \\ \times\ 2 \\ \hline \end{array}$$ ....

$$\begin{array}{r} 2 \\ \times\ 2 \\ \hline \end{array}$$ ....

$$\begin{array}{r} 8 \\ \times\ 4 \\ \hline \end{array}$$ ....

$$\begin{array}{r} 8 \\ \times\ 8 \\ \hline \end{array}$$ ....

$$9 \times 8$$

$$8 \times 5$$

$$9 \times 3$$

$$4 \times 6$$

$$9 \times 9$$

$$3 \times 8$$

$$10 \times 5$$

$$5 \times 1$$

$$3 \times 9$$

$$6 \times 6$$

$$2 \times 10$$

$$2 \times 9$$

$$8 \times 10$$

$$5 \times 2$$

$$4 \times 7$$

1 × 1	3 × 10	1 × 6
9 × 7	2 × 6	1 × 3
3 × 1	4 × 9	1 × 2
5 × 9	2 × 5	1 × 4
7 × 5	1 × 9	7 × 6

5 × 4	8 × 6	9 × 4
....	....	....
4 × 2	2 × 6	7 × 2
....	....	....
5 × 8	6 × 5	5 × 7
....	....	....
7 × 9	7 × 6	7 × 7
....	....	....
3 × 7	8 × 4	6 × 9
....	....	....

multiplication

8	6	8
× 9	× 7	× 10

4	7	8
× 3	× 1	× 3

3	3	7
× 2	× 4	× 5

3	2	3
× 8	× 9	× 3

6	5	8
× 1	× 3	× 8

10 × 6	2 × 7	2 × 5
4 × 6	5 × 9	3 × 5
7 × 3	7 × 10	4 × 5
3 × 9	8 × 1	9 × 8
7 × 4	2 × 2	4 × 4

6 × 8	6 × 2	3 × 1
4 × 7	6 × 6	5 × 5
5 × 6	9 × 7	9 × 3
2 × 4	8 × 7	1 × 7
4 × 8	6 × 4	9 × 9

9 × 2	5 × 10	6 × 10
1 × 6	4 × 10	5 × 2
1 × 2	1 × 9	1 × 3
10 × 4	9 × 5	10 × 7
8 × 5	10 × 2	2 × 3

multiplication

4 × 1	4 × 9	3 × 6
....	....	....
9 × 6	10 × 5	2 × 10
....	....	....
10 × 1	10 × 8	2 × 8
....	....	....
5 × 1	1 × 8	1 × 1
....	....	....
6 × 3	1 × 10	8 × 2
....	....	....

$$\begin{array}{r} 9 \\ \times\ 5 \\ \hline \end{array}$$
....

$$\begin{array}{r} 1 \\ \times\ 6 \\ \hline \end{array}$$
....

$$\begin{array}{r} 7 \\ \times\ 2 \\ \hline \end{array}$$
....

$$\begin{array}{r} 8 \\ \times\ 2 \\ \hline \end{array}$$
....

$$\begin{array}{r} 2 \\ \times\ 9 \\ \hline \end{array}$$
....

$$\begin{array}{r} 9 \\ \times\ 6 \\ \hline \end{array}$$
....

$$\begin{array}{r} 5 \\ \times\ 2 \\ \hline \end{array}$$
....

$$\begin{array}{r} 3 \\ \times\ 5 \\ \hline \end{array}$$
....

$$\begin{array}{r} 8 \\ \times\ 9 \\ \hline \end{array}$$
....

$$\begin{array}{r} 6 \\ \times\ 10 \\ \hline \end{array}$$
....

$$\begin{array}{r} 8 \\ \times\ 7 \\ \hline \end{array}$$
....

$$\begin{array}{r} 8 \\ \times\ 1 \\ \hline \end{array}$$
....

$$\begin{array}{r} 9 \\ \times\ 7 \\ \hline \end{array}$$
....

$$\begin{array}{r} 10 \\ \times\ 10 \\ \hline \end{array}$$
....

$$\begin{array}{r} 1 \\ \times\ 7 \\ \hline \end{array}$$
....

multiplication

7 × 7	1 × 9	5 × 3
3 × 4	6 × 9	2 × 2
2 × 3	9 × 4	10 × 2
3 × 9	7 × 5	8 × 6
3 × 3	8 × 3	5 × 6

7	4	8
× 9	× 6	× 4
....	....	....
5	6	2
× 7	× 7	× 8
....	....	....
4	3	2
× 2	× 1	× 4
....	....	....
7	2	9
× 10	× 5	× 3
....	....	....
6	10	4
× 3	× 4	× 7
....	....	....

6 × 6	5 × 1	7 × 6
4 × 5	9 × 10	7 × 3
10 × 7	9 × 9	6 × 5
3 × 6	2 × 7	7 × 4
3 × 8	5 × 4	4 × 4

5 × 9	9 × 1	3 × 10
6 × 2	4 × 8	2 × 10
5 × 8	4 × 3	1 × 2
8 × 8	1 × 4	4 × 9
2 × 6	3 × 2	9 × 2

$$\begin{array}{r} 4 \\ \times\ 1 \\ \hline \end{array}$$

$$\begin{array}{r} 10 \\ \times\ 6 \\ \hline \end{array}$$

$$\begin{array}{r} 1 \\ \times\ 8 \\ \hline \end{array}$$

$$\begin{array}{r} 1 \\ \times\ 5 \\ \hline \end{array}$$

$$\begin{array}{r} 2 \\ \times\ 1 \\ \hline \end{array}$$

$$\begin{array}{r} 7 \\ \times\ 8 \\ \hline \end{array}$$

$$\begin{array}{r} 3 \\ \times\ 7 \\ \hline \end{array}$$

$$\begin{array}{r} 8 \\ \times\ 5 \\ \hline \end{array}$$

$$\begin{array}{r} 8 \\ \times\ 10 \\ \hline \end{array}$$

$$\begin{array}{r} 4 \\ \times\ 10 \\ \hline \end{array}$$

$$\begin{array}{r} 6 \\ \times\ 8 \\ \hline \end{array}$$

$$\begin{array}{r} 10 \\ \times\ 1 \\ \hline \end{array}$$

$$\begin{array}{r} 9 \\ \times\ 8 \\ \hline \end{array}$$

$$\begin{array}{r} 1 \\ \times\ 10 \\ \hline \end{array}$$

$$\begin{array}{r} 5 \\ \times\ 5 \\ \hline \end{array}$$

6	8	7
× 4	× 2	× 9

4	4	3
× 6	× 8	× 4

6	5	7
× 7	× 4	× 4

6	10	3
× 2	× 7	× 9

8	2	5
× 9	× 8	× 3

$$7 \times 5$$

$$8 \times 10$$

$$7 \times 6$$

$$1 \times 5$$

$$6 \times 6$$

$$8 \times 4$$

$$5 \times 6$$

$$6 \times 9$$

$$2 \times 6$$

$$7 \times 7$$

$$4 \times 3$$

$$10 \times 9$$

$$3 \times 5$$

$$4 \times 1$$

$$8 \times 5$$

7	4	2
× 1	× 10	× 10
....	....	....
9	7	6
× 8	× 10	× 10
....	....	....
7	3	1
× 3	× 1	× 9
....	....	....
2	7	3
× 7	× 2	× 7
....	....	....
7	4	5
× 8	× 2	× 10
....	....	....

$$\begin{array}{r} 9 \\ \times\ 4 \\ \hline \end{array}$$

$$\begin{array}{r} 5 \\ \times\ 8 \\ \hline \end{array}$$

$$\begin{array}{r} 5 \\ \times\ 2 \\ \hline \end{array}$$

$$\begin{array}{r} 8 \\ \times\ 1 \\ \hline \end{array}$$

$$\begin{array}{r} 6 \\ \times\ 3 \\ \hline \end{array}$$

$$\begin{array}{r} 3 \\ \times\ 2 \\ \hline \end{array}$$

$$\begin{array}{r} 9 \\ \times\ 7 \\ \hline \end{array}$$

$$\begin{array}{r} 8 \\ \times\ 3 \\ \hline \end{array}$$

$$\begin{array}{r} 1 \\ \times\ 7 \\ \hline \end{array}$$

$$\begin{array}{r} 4 \\ \times\ 9 \\ \hline \end{array}$$

$$\begin{array}{r} 9 \\ \times\ 6 \\ \hline \end{array}$$

$$\begin{array}{r} 9 \\ \times\ 9 \\ \hline \end{array}$$

$$\begin{array}{r} 10 \\ \times\ 4 \\ \hline \end{array}$$

$$\begin{array}{r} 6 \\ \times\ 5 \\ \hline \end{array}$$

$$\begin{array}{r} 2 \\ \times\ 3 \\ \hline \end{array}$$

multiplication

$$8 \times 8$$

....

$$1 \times 8$$

....

$$3 \times 8$$

....

$$2 \times 5$$

....

$$1 \times 3$$

....

$$4 \times 7$$

....

$$3 \times 10$$

....

$$9 \times 3$$

....

$$5 \times 7$$

....

$$9 \times 5$$

....

$$10 \times 5$$

....

$$2 \times 1$$

....

$$10 \times 2$$

....

$$6 \times 8$$

....

$$9 \times 10$$

....

5 × 9	2 × 2	3 × 3
5 × 5	2 × 9	4 × 4
6 × 1	2 × 4	10 × 1
8 × 7	5 × 1	1 × 2
3 × 6	1 × 10	8 × 6

multiplication

$$
\begin{array}{r}
2 \\
\times\ 7 \\
\hline
\end{array}
\qquad
\begin{array}{r}
2 \\
\times\ 6 \\
\hline
\end{array}
\qquad
\begin{array}{r}
9 \\
\times\ 8 \\
\hline
\end{array}
$$

....

$$
\begin{array}{r}
6 \\
\times\ 4 \\
\hline
\end{array}
\qquad
\begin{array}{r}
10 \\
\times\ 10 \\
\hline
\end{array}
\qquad
\begin{array}{r}
8 \\
\times\ 3 \\
\hline
\end{array}
$$

....

$$
\begin{array}{r}
9 \\
\times\ 4 \\
\hline
\end{array}
\qquad
\begin{array}{r}
7 \\
\times\ 3 \\
\hline
\end{array}
\qquad
\begin{array}{r}
2 \\
\times\ 5 \\
\hline
\end{array}
$$

....

$$
\begin{array}{r}
10 \\
\times\ 6 \\
\hline
\end{array}
\qquad
\begin{array}{r}
2 \\
\times\ 9 \\
\hline
\end{array}
\qquad
\begin{array}{r}
9 \\
\times\ 9 \\
\hline
\end{array}
$$

....

$$
\begin{array}{r}
2 \\
\times\ 3 \\
\hline
\end{array}
\qquad
\begin{array}{r}
5 \\
\times\ 6 \\
\hline
\end{array}
\qquad
\begin{array}{r}
5 \\
\times\ 2 \\
\hline
\end{array}
$$

....

10	7	8
× 1	× 7	× 6

4	9	3
× 5	× 2	× 2

6	4	2
× 6	× 7	× 4

7	9	3
× 6	× 7	× 8

6	8	6
× 3	× 9	× 5

multiplication

5 × 4	7 × 10	7 × 8
2 × 10	3 × 9	4 × 2
4 × 4	3 × 3	3 × 6
1 × 7	5 × 8	9 × 3
9 × 5	9 × 10	2 × 8

multiplication

$$4 \times 3$$

$$7 \times 2$$

$$7 \times 4$$

$$5 \times 5$$

$$7 \times 5$$

$$5 \times 10$$

$$5 \times 7$$

$$6 \times 9$$

$$6 \times 8$$

$$8 \times 2$$

$$10 \times 8$$

$$10 \times 5$$

$$4 \times 6$$

$$2 \times 2$$

$$10 \times 9$$

$$1 \times 9$$

$$8 \times 4$$

$$8 \times 5$$

$$3 \times 7$$

$$10 \times 4$$

$$8 \times 1$$

$$1 \times 6$$

$$1 \times 5$$

$$9 \times 6$$

$$6 \times 7$$

$$10 \times 7$$

$$8 \times 10$$

$$8 \times 7$$

$$6 \times 1$$

$$3 \times 4$$

$$
\begin{array}{r} 4 \\ \times\ 8 \\ \hline \end{array}
\qquad
\begin{array}{r} 8 \\ \times\ 8 \\ \hline \end{array}
\qquad
\begin{array}{r} 6 \\ \times\ 2 \\ \hline \end{array}
$$

$$
\begin{array}{r} 4 \\ \times\ 1 \\ \hline \end{array}
\qquad
\begin{array}{r} 1 \\ \times\ 3 \\ \hline \end{array}
\qquad
\begin{array}{r} 6 \\ \times\ 10 \\ \hline \end{array}
$$

$$
\begin{array}{r} 1 \\ \times\ 2 \\ \hline \end{array}
\qquad
\begin{array}{r} 10 \\ \times\ 3 \\ \hline \end{array}
\qquad
\begin{array}{r} 7 \\ \times\ 9 \\ \hline \end{array}
$$

$$
\begin{array}{r} 5 \\ \times\ 1 \\ \hline \end{array}
\qquad
\begin{array}{r} 2 \\ \times\ 1 \\ \hline \end{array}
\qquad
\begin{array}{r} 3 \\ \times\ 1 \\ \hline \end{array}
$$

$$
\begin{array}{r} 4 \\ \times\ 10 \\ \hline \end{array}
\qquad
\begin{array}{r} 5 \\ \times\ 9 \\ \hline \end{array}
\qquad
\begin{array}{r} 1 \\ \times\ 4 \\ \hline \end{array}
$$